PLAYGROUND EDUCATION

CREATED and WRITTEN by

R. TOBIAS PITTMAN

THE EDUCATIONAL PANDA

PUGSLEY PANDA

ILLUSTRATIONS by

William Chislum

EDITED by

Gloria Miles

Playground Education and Entertainment
www.pugsleypanda.com | email: pugsleypanda@gmail.com

Look at those clouds.

Hello Pugsley, do you have an umbrella? I think it is going to rain.
Hello Renee, I do not have an umbrella. May I share your umbrella with you?

Sure, you can share my umbrella.
Hey Pugsley,
where does rain come from?
That is a good question Renee,
maybe Science Scott can explain
where rain comes from.

Hello everyone, did someone call my name?
Hello Science Scott, we were talking about the rain. You know all about Science, maybe you can explain to us where rain comes from.

Listen carefully, I will explain where rain come from.
Great, I am excited!

Actually all the water we see and use including rain, has been around for millions of years. The water keeps going up to the clouds and back to the ground, up to the clouds and back to the ground. This is called "The Water Cycle".
Wow, that sounds interesting. How does that work?

The cycle is made up of four main parts; Evaporation (Transpiration), Condensation, Precipitation, and Collection.
Continue, Science Scott.

The sun's heat evaporates water from the oceans, rivers, and lakes, into the atmosphere to form clouds.
You are very knowledgeable, Science Scott.

The next step in the water cycle process is Transpiration.
How does Transpiration work, Science Scott?

Transpiration is the process by which plants lose water out of their leaves.
Transpiration gives evaporation a bit of a hand in getting the water back up into the air.
This is great information!

Water vapor in the air gets cold and changes back into liquid, forming clouds. This is called condensation.
I think we're getting it.
Now I know what condensation is.

The same thing happens at home, when you pour a glass of cold water on hot day. Water forms on the outside of the glass.
Some people think the water somehow magically leaked through the glass, but it actually came from the air.
Water vapor in the warm air, turns back into liquid when it touches the cold glass.
This is terrific!

Precipitation is the next step in the water cycle. Precipitation happens when so much water has condensed, that the air cannot hold it anymore. The clouds get heavy and water falls back to the earth in the form of rain, hail, sleet or snow.
Wow, that's incredible. I need my umbrella again.

Collection is the next step. When water falls back to earth as precipitation it may fall back in the oceans, lakes or rivers and the water is collected.
This is awesome, Science Scott!
If the water ends up on land, it can soak into the earth and become part of the ground water that plants and animals use to drink. This is where the cycle starts all over again.

Do you guys understand the water cycle now?
The explanation of the water cycle was very interesting and educational, Science Scott.
How can I remember the steps?

Repeat after me, "Every child picks cherries".
Every child picks cherries.

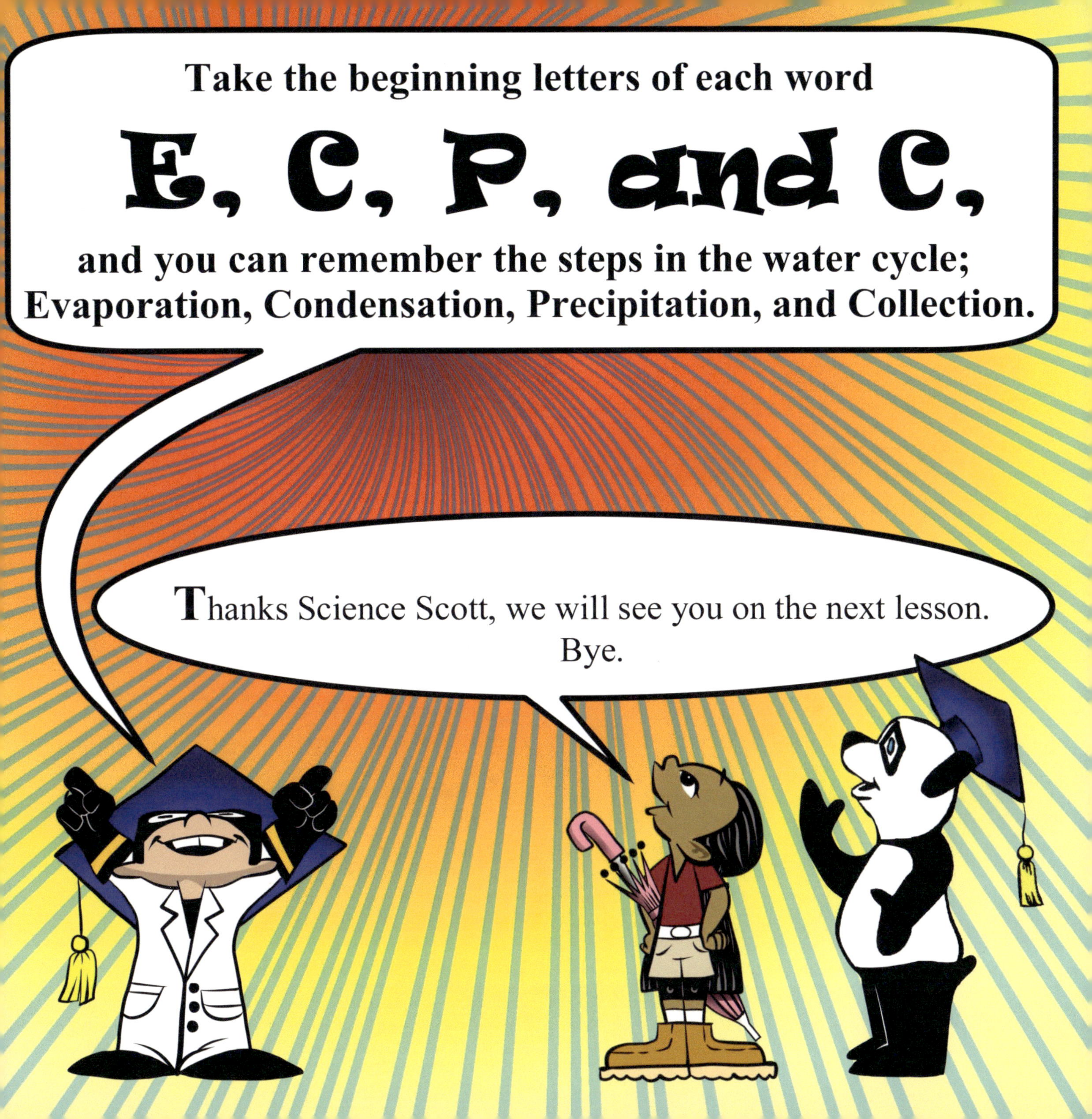
Take the beginning letters of each word
E, C, P, and C,
and you can remember the steps in the water cycle; Evaporation, Condensation, Precipitation, and Collection.
Thanks Science Scott, we will see you on the next lesson. Bye.

Every Child Picks Cherries

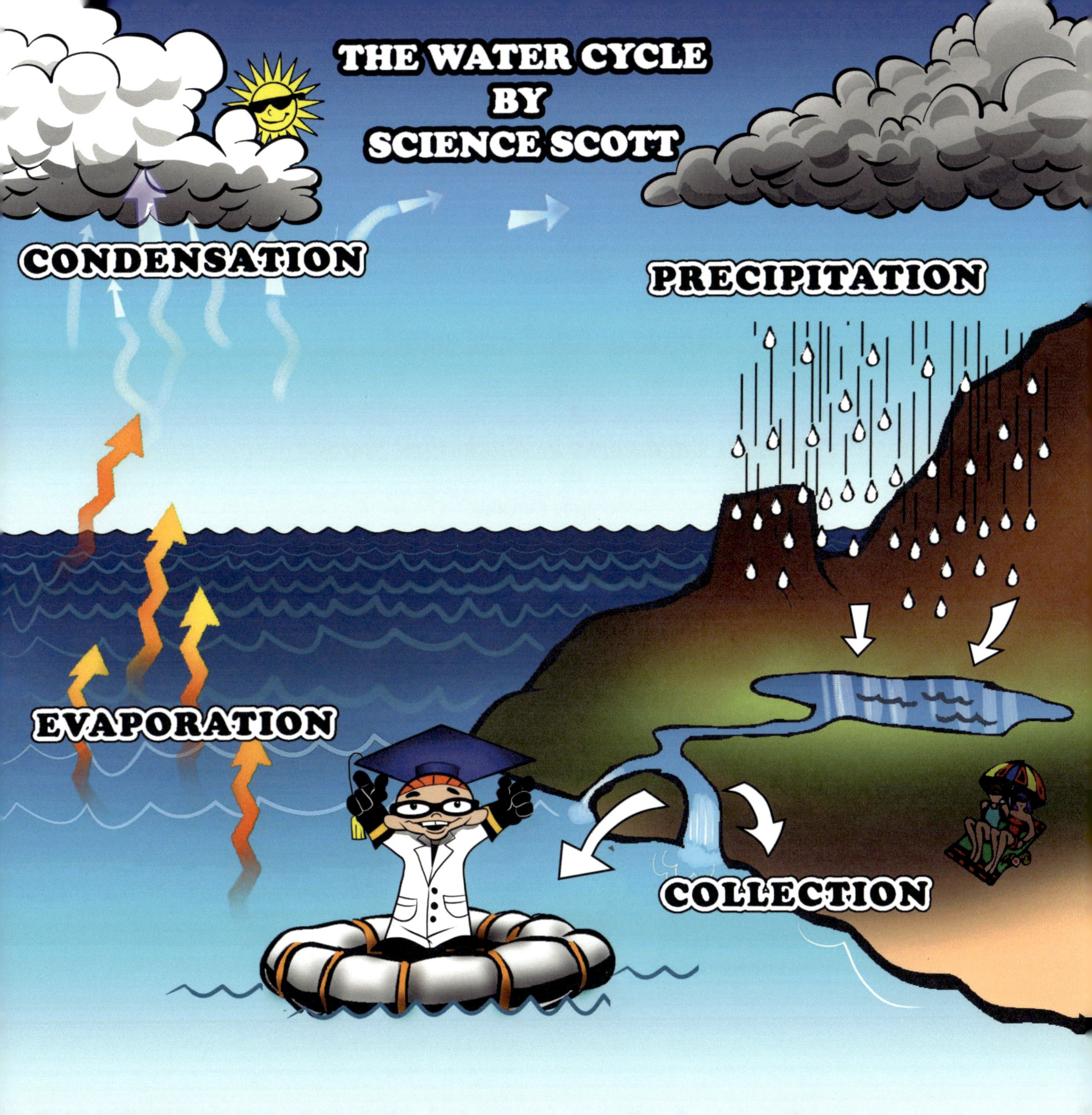
THE WATER CYCLE
BY
SCIENCE SCOTT
CONDENSATION
PRECIPITATION
EVAPORATION
COLLECTION

ENJOY THESE OTHER CHILDREN'S BOOKS FROM

PLAYGROUND EDUCATION

"No Bullying"
By
Safety Sam

"Wellington Hill Playground"
By
The Educational Kidz

"Famous Landmarks in The United States"
By
Geography Gerald

"Seasonal Safety Tips for Children"
By
Safety Sam

PUGSLEY PANDA

Made in the USA
Las Vegas, NV
07 November 2026

57650025R00017